KS1 Success

Age 5-7

Maths

Trevor Dixon

Practice Workbook

Contents

Statistics

Mixed Practice Questions

Glossary

Answers

Set in a pull-out booklet in the centre of the book

1 Write the next three numbers in each sequence.

24 26 28 30

a. 24 26 28 30 32 34 .... 38 **(1 mark)**

b. 18 21 24 27 30 .... **(1 mark)**

c. 30 35 40 45 50 .... **(1 mark)**

d. 20 30 40 50 60 .... **(1 mark)**

2 Write the next three numbers in each sequence.

a. 28 26 24 22 20 .... **(1 mark)**

b. 21 18 15 12 9 .... **(1 mark)**

c. 40 35 30 25 20 .... **(1 mark)**

d. 95 85 75 65 55 .... **(1 mark)**

3 Write the missing numbers in each sequence.

a. 31 33 35 .... 39 41 .... 45 .... **(1 mark)**

b. 48 45 .... 39 .... 33 30 .... **(1 mark)**

c. 29 34 39 .... 49 54 .... 69 **(1 mark)**

d. 82 72 .... 52 .... 32 22 12 .... **(1 mark)**

4 How do the numbers in this pattern change? **(1 mark)**

64 59 54 49 44 39

....

5 Count from 31 to 81.

How many tens did you count? **(1 mark)**

....

6 Here is part of a number line.

Write 52 in the correct place. **(1 mark)**

			42	44	46				

7 Here is part of a number line.

Write 57 in the correct place. **(1 mark)**

				66	69	72			

8 Here is part of a 100 square.

	45	46	47	48	49	

a. Write 54 in the correct place. **(1 mark)**

b. Write 70 in the correct place. **(1 mark)**

9 Start at 56. Count on in fives.

Write the next 3 numbers. **(1 mark)**

....

Top tip! Practise counting in twos, fives and tens. This will help you answer these kinds of questions.

Total $\frac{}{19}$

1 Write the missing numbers.

a. 67 = ... tens and 7 units **(1 mark)**

b. 48 = ... tens and .... units **(1 mark)**

c. 83 = ... units and .... tens **(1 mark)**

d. .... = 7 tens and 5 units **(1 mark)**

e. .... = 4 units and 9 tens **(1 mark)**

2 Here are two place value cards.

What number do they make? **(1 mark)**

....

3 Here are two different place value cards.

What number do they make? **(1 mark)**

....

4 What number does this abacus show? **(1 mark)**

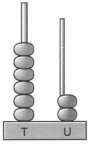

T   U   ....

Ask your child how many tens and units are in 2-digit numbers. Use numbers from your day-to-day life, such as house numbers, page numbers and numbers on a menu.

Parent tip!

5 Write the answers. **(3 marks)**

   a. 37 + 10 = ...

   b. 74 + 20 = ...

   c. 28 + 40 = ...

6 Here is part of a 100 square.

Write the missing numbers. **(4 marks)**

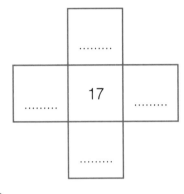

7 Write the answers. **(3 marks)**

   a. 54 − 10 = ...

   b. 87 − 50 = ...

   c. 49 − 40 = ...

8 Here are some digit cards.

   a. Write the largest 2-digit number you can make with 7 as the tens. **(1 mark)**

   ....

   b. Write the smallest 2-digit number you can make with 8 as the units. **(1 mark)**

   ....

9 Tom is thinking of a number. It has 4 tens and 8 units.

   What is Tom's number? .... **(1 mark)**

Total $\dfrac{}{21}$

1 Here are some digit cards.

Use the cards to make:

a. the largest 2-digit number. .... **(1 mark)**

b. the smallest, odd, 2-digit number. .... **(1 mark)**

c. the largest, even, 2-digit number. .... **(1 mark)**

2 This stands for 10:

This stands for 1:

What number does this stand for? **(1 mark)**

....

3 This stands for 10:

This stands for 1:

What number does this stand for? **(1 mark)**

....

④ What numbers are the arrows pointing to?

a. 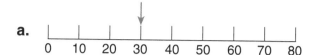 .... **(1 mark)**

b.  .... **(1 mark)**

c.  .... **(1 mark)**

d.  .... **(1 mark)**

⑤ On each number line, draw an arrow to show the number.

a. Show 70. **(1 mark)**

b. Show 10. **(1 mark)**

c. Show 30. **(1 mark)**

d. Show 15. **(1 mark)**

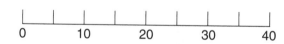

Total — 13

1 Write these numbers using digits.

a. thirty-five .... (1 mark)

b. seventy-six .... (1 mark)

c. fifty .... (1 mark)

d. eighty-three .... (1 mark)

2 Write these numbers as words.

a. 20 .... (1 mark)

b. 16 .... (1 mark)

c. 92 .... (1 mark)

d. 64 .... (1 mark)

3 Put these numbers in order. Start with the smallest. (1 mark)

56   43   19   70

....

....

....

....

4 Put these numbers in order. Start with the largest. (1 mark)

61   75   68   17

....

....

....

....

5 Insert <, > or = to compare these numbers. **(6 marks)**

a. 67 ◯ 56

b. 78 ◯ 19

c. 63 ◯ 67

d. 42 ◯ 42

e. 19 ◯ 91

f. 39 ◯ 31

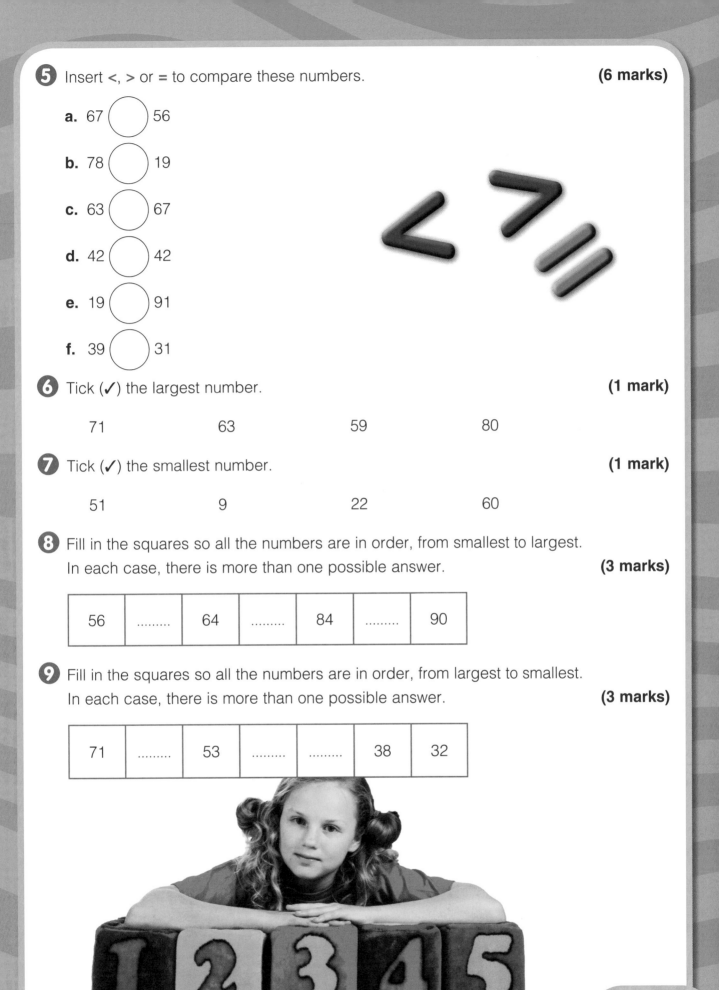

6 Tick (✓) the largest number. **(1 mark)**

71   63   59   80

7 Tick (✓) the smallest number. **(1 mark)**

51   9   22   60

8 Fill in the squares so all the numbers are in order, from smallest to largest.
In each case, there is more than one possible answer. **(3 marks)**

56	....	64	....	84	....	90

9 Fill in the squares so all the numbers are in order, from largest to smallest.
In each case, there is more than one possible answer. **(3 marks)**

71	....	53	....	....	38	32

Total ____
24

1 Write the answers.

  a. 15 + 6 = ... **(1 mark)**

  b. 56 + 5 = ... **(1 mark)**

  c. 38 + 6 = ... **(1 mark)**

  d. 75 + 7 = ... **(1 mark)**

2 Write the answers.

  a. 22 − 5 = ... **(1 mark)**

  b. 41 − 4 = ... **(1 mark)**

  c. 55 − 6 = ... **(1 mark)**

  d. 74 − 8 = ... **(1 mark)**

3 This is an addition grid.

  When you move one column right, you add 2.

  When you move one column down, you add 5.

  Fill in the empty squares. **(3 marks)**

Add 2 →

Add 5 ↓

12	14	....		
	19	21	....	
		....	28	
		31	....	
			38	....

4 Add these numbers.

    a. 6 + 4 + 3 = ...   **(1 mark)**

    b. 7 + 5 + 8 = ...   **(1 mark)**

    c. 5 + 2 + 9 = ...   **(1 mark)**

    d. 4 + 8 + 9 = ...   **(1 mark)**

5 Write the missing numbers.

    a. 34 + ... = 40   **(1 mark)**

    b. 24 − .... = 19   **(1 mark)**

    c. .... + 84 = 88   **(1 mark)**

    d. .... − 6 = 18   **(1 mark)**

    e. 7 + 4 + ... = 16   **(1 mark)**

    f. 8 + ... + 4 = 17   **(1 mark)**

6 Tick (✔) the correct number sentences. Write a cross (✗) if they are incorrect.

    a. 4 + 26 = 26 + 4   □   **(1 mark)**

    b. 25 − 3 = 3 − 25   □   **(1 mark)**

    c. 35 + 9 = 9 + 35   □   **(1 mark)**

    d. 56 − 8 = 8 − 56   □   **(1 mark)**

    e. 4 + 5 + 8 = 8 + 4 + 5   □   **(1 mark)**

7 Write a pair of numbers that total 19.   **(1 mark)**

.... + ... = 19

8 Write three numbers that total 12.   **(1 mark)**

.... + ... + ... = 12

Top tip! Work out sums on paper if you need to. You do not have to work everything out in your head.

Total —— 28

1 Write the answers. **(6 marks)**

a. 6 × 2 = ...

d. 8 × 2 = ...

b. 4 × 5 = ...

e. 3 × 5 = ...

c. 7 × 10 = ...

f. 9 × 5 = ...

2 Write the missing numbers. **(6 marks)**

a. 5 × .... = 10

d. .... × 5 = 30

b. 8 × .... = 80

e. .... × 2 = 18

c. 5 × .... = 25

f. .... × 10 = 40

3 This is an array. ☆ ☆ ☆ ☆ ☆ ☆ ☆ ☆ ☆ ☆
☆ ☆ ☆ ☆ ☆ ☆ ☆ ☆ ☆ ☆

Which multiplication fact does it show? **(1 mark)**

.... × .... = ...

4 Write the answers. **(6 marks)**

a. 14 ÷ 2 = ...

d. 25 ÷ 5 = ...

b. 60 ÷ 10 = ...

e. 8 ÷ 2 = ...

c. 35 ÷ 5 = ...

f. 40 ÷ 5 = ...

5 Write the multiplication sum that matches each addition.

a. 6 + 6 + 6 + 6 + 6 =

.... × .... = ... **(1 mark)**

b. 2 + 2 + 2 + 2 + 2 + 2 + 2 + 2 + 2 + 2 =

.... × .... = ... **(1 mark)**

c. 8 + 8 =

.... × .... = ... **(1 mark)**

d. 2 + 2 + 2 + 2 + 2 =

.... × .... = ... **(1 mark)**

6 This is a multiplication grid.

Fill in the missing numbers.

(3 marks)

×	2	5	10
3	6	....	....
7	....	35	....
8	....	....	80

7 Here is a different multiplication grid.

Fill in the missing numbers.

(3 marks)

×	2	5	10
....	8	20	....
....	....	25	50
....	18	....	90

8 Use the numbers 2, 8 and 16 to make four different facts.

a. .... × .... = ... **(1 mark)**

b. .... × .... = ... **(1 mark)**

c. .... ÷ .... = ... **(1 mark)**

d. .... ÷ .... = ... **(1 mark)**

9 Tick (✓) the correct number sentence(s). Insert a cross (✗) if it is incorrect.

a. $8 \times 5 = 5 \times 8$ ☐ **(1 mark)**

b. $18 \div 2 = 2 \div 18$ ☐ **(1 mark)**

Total $\dfrac{}{35}$

① Work out:

a. 55 + 34 = ... **(1 mark)**   **e.** 57 + 69 = ... **(1 mark)**

b. 67 − 35 = ... **(1 mark)**   **f.** 45 − 18 = ... **(1 mark)**

c. 68 − 30 = ... **(1 mark)**   **g.** 70 − 37 = ... **(1 mark)**

d. 28 + 42 = ... **(1 mark)**   **h.** 67 + 45 = ... **(1 mark)**

② Write the missing numbers in these column calculations.

a.   **(1 mark)**

	7	....
+	....	5
	9	8

b.   **(1 mark)**

	8	....
−	....	8
	5	6

③ Circle the odd numbers.   **(2 marks)**

45   80   51   68   13   96   34   27

④ Solve these problems. Show your working.

a. Yasmin has a tub of 40 counters. She takes 27 out.
How many counters are left?   **(2 marks)**

.... counters

b. Polly has 28 stickers. She buys another 15.
How many stickers does she have altogether?   **(2 marks)**

.... stickers

c. Nishi has 5 bags. There are 6 toys in each bag.

How many toys does Nishi have altogether? **(2 marks)**

.... toys

d. Sam has 20 crayons. He puts them into 5 piles.

How many crayons are in each pile? **(2 marks)**

.... crayons

e. Kassim has 35 chocolates. He shares them between 5 friends.

They each get the same number of chocolates.

How many chocolates does each friend get? **(2 marks)**

.... chocolates

Encourage your child to check their work, to help avoid careless mistakes.

Parent tip!

Total $\frac{}{22}$

Fractions

① Draw lines from the fractions to the fraction names. **(2 marks)**
The first one has been done for you.

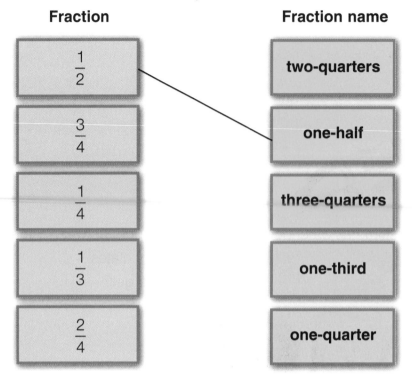

② What fraction of each shape is shaded blue?
Write your answers as fractions.

a. = ... **(1 mark)**   c. = ... **(1 mark)**

b. = ... **(1 mark)**   d. = ... **(1 mark)**

③ Here are some green and pink counters.

a. What fraction of the counters is green?   .... **(1 mark)**

b. What fraction of the counters is pink?   .... **(1 mark)**

④ Tick (✓) the two shapes that have $\frac{1}{4}$ shaded orange.

(2 marks)

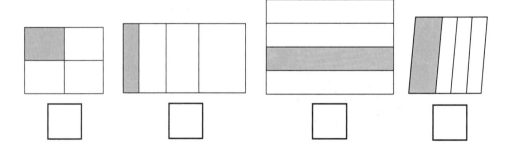

5 Tick (✓) the two bags that have half of the counters shaded purple.

(2 marks)

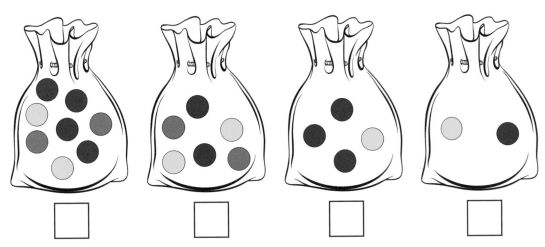

☐ ☐ ☐ ☐

6 Tick (✓) the shape that has $\frac{3}{4}$ shaded red.

(1 mark)

☐ ☐ ☐ ☐

7 Tick (✓) the row that is correct.

(1 mark)

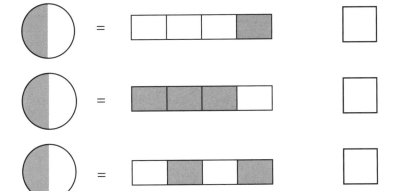

 When your child is finding a fraction, the quick way is to divide by the bottom number. Remember, though, that your child will need to multiply by the top number if it is larger than 1.

Parent tip!

Total $\frac{}{14}$

1 Shade $\frac{1}{2}$ of each shape.

a. **(1 mark)**

b. **(1 mark)**

c. **(1 mark)**

d. **(1 mark)**

2 a. Shade $\frac{1}{3}$ of this shape. **(1 mark)**

b. Shade $\frac{3}{4}$ of this shape.  **(1 mark)**

c. Shade $\frac{1}{4}$ of this shape. 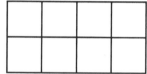 **(1 mark)**

d. Shade $\frac{1}{3}$ of this shape. 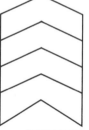 **(1 mark)**

3 Below are 12 counters.

Shade $\frac{1}{4}$ of them. **(1 mark)**

4 Below are 8 counters.

Shade $\frac{3}{4}$ of them. **(1 mark)**

5 Work out:

a. $\frac{1}{2}$ of 12 cm = ... cm **(1 mark)**   **d.** $\frac{3}{4}$ of 20 kg = ... kg **(1 mark)**

b. $\frac{1}{3}$ of £6 = £ .... **(1 mark)**   **e.** $\frac{1}{3}$ of 15 m = ... m **(1 mark)**

c. $\frac{1}{4}$ of 20 kg = ... kg **(1 mark)**   **f.** $\frac{3}{4}$ of £12 = £ .... **(1 mark)**

6 Solve these problems. Show your working.

a. Ben has £16. He spends $\frac{1}{2}$ of the money.

How much does he spend? **(2 marks)**

£ ....

b. Fay has 8 kg of potatoes. She uses $\frac{1}{4}$ of the potatoes.

How many kg of potatoes has she used? **(2 marks)**

.... kg

c. Tariq's book has 30 pages. He has read $\frac{1}{3}$ of the pages.

How many pages has he read? **(2 marks)**

.... pages

d. Fatima has 12 toy bears. Three-quarters of the bears are brown.

How many of the toy bears are brown? **(2 marks)**

.... toy bears

7 Dan has £20. He spends £5.

a. What fraction of the money has Dan spent? .... **(1 mark)**

b. What fraction of the money does he still have? .... **(1 mark)**

Total $\frac{}{26}$

Measurement

Units and scales

❶ Draw lines to join the equipment to what it would be used for.
The first one has been done for you. **(2 marks)**

Equipment	What the equipment is used for
rulers	measure mass
scales	measure capacity
thermometers	measure length
measuring jugs	
tapes	measure temperature

2000
1500
500
1000

❷ Tick (✓) the unit for measuring weight. **(1 mark)**

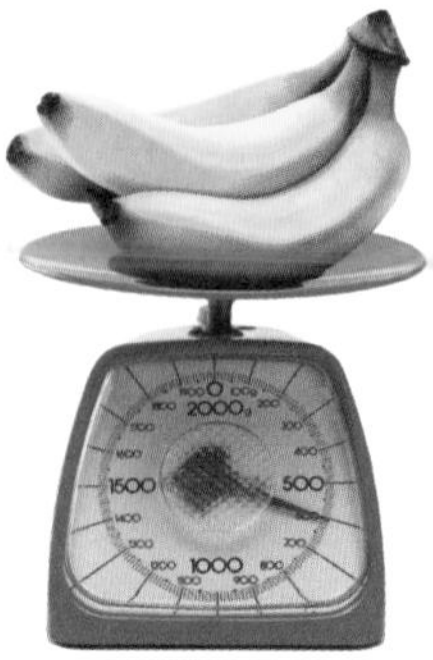

kilogram ☐ litre ☐ centimetre ☐ degree ☐

❸ Tick (✓) the unit for measuring capacity. **(1 mark)**

degree ☐ metre ☐ litre ☐ gram ☐

❹ Tick (✓) the unit for measuring length. **(1 mark)**

millilitre ☐ centimetre ☐ gram ☐ litre ☐

❺ How long is this line? **(1 mark)**

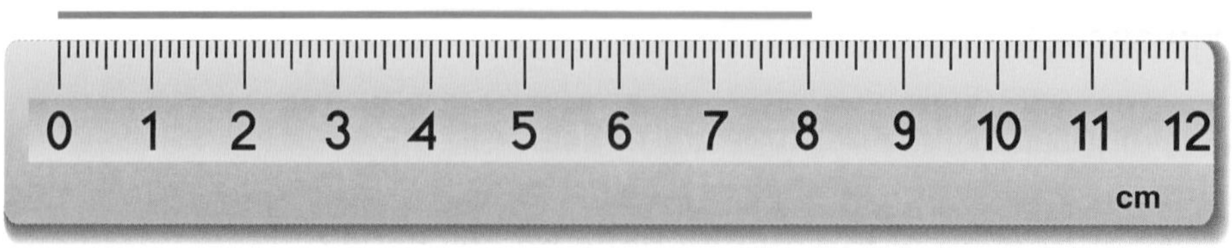

0 1 2 3 4 5 6 7 8 9 10 11 12
cm

.........cm

6 What is the temperature? **(1 mark)**

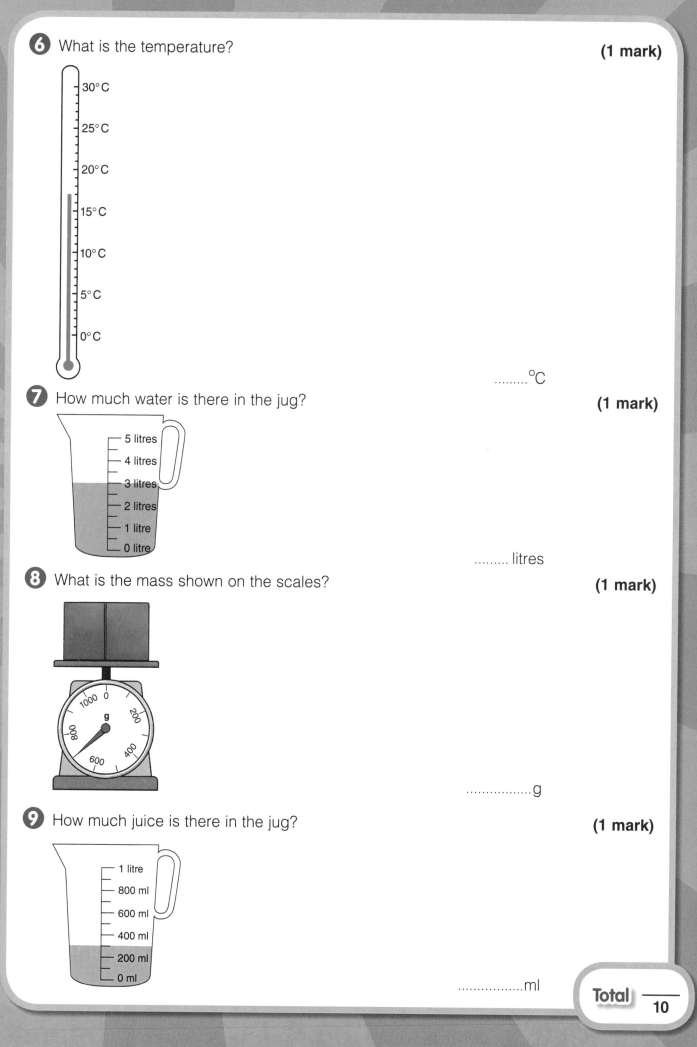

.........°C

7 How much water is there in the jug? **(1 mark)**

.... litres

8 What is the mass shown on the scales? **(1 mark)**

.................g

9 How much juice is there in the jug? **(1 mark)**

.................ml

Total $\dfrac{}{10}$

1 List these animals in order of height, shortest first. **(1 mark)**

horse　　　**rabbit**　　　**elephant**　　　**dog**

....

2 List these containers in order of size, largest first. **(1 mark)**

bucket　　　**cup**　　　**bath**　　　**bottle**

....

3 List these vehicles in order of mass, lightest first. **(1 mark)**

lorry　　　**car**　　　**bike**　　　**minibus**

....

4 Tick (✓) the best weight for an apple. **(1 mark)**

3 litres　　　　1 metre　　　　5 kilograms　　　　150 grams

☐　　　　☐　　　　☐　　　　☐

5 Tick (✓) the best capacity for a mug. **(1 mark)**

5 litres　　　300 millilitres　　　5 centimetres　　　40°C

☐　　　　☐　　　　☐　　　　☐

6 Insert <, > or = to compare these units.

a. 2 kg ◯ 3 kg **(1 mark)**　　**d.** 30 cm ◯ 30 cm **(1 mark)**

b. 5 m ◯ 10 m **(1 mark)**　　**e.** 150 cm ◯ 75 cm **(1 mark)**

c. 900 ml ◯ 800 ml **(1 mark)**　　**f.** 3000 g ◯ 8000 g **(1 mark)**

7 Insert <, > or = to compare these units.

a. 1 kg $\bigcirc$ 1000 g **(1 mark)**

b. 5 litres $\bigcirc$ 2000 ml **(1 mark)**

c. 1 m $\bigcirc$ 75 cm **(1 mark)**

d. 3 m $\bigcirc$ 300 cm **(1 mark)**

e. 3000 g $\bigcirc$ 2 kg **(1 mark)**

f. 4 litres $\bigcirc$ 4000 ml **(1 mark)**

8 Solve these problems.

a. Milly has two dogs. Lola weighs 5 kg and Fluffy weighs $4\frac{1}{2}$ kg.
Put a circle around the heavier dog. **(1 mark)**

Lola   **Fluffy**

b. Magnus buys two bottles of drink.
There is a 2 litre bottle of lemonade and a 1 litre bottle of juice.
Put a circle around the bottle that holds more. **(1 mark)**

1 litre   **2 litres**

c. Sally measures two bits of ribbon.
The blue ribbon is 2 metres long. The red ribbon is 100 centimetres long.
Put a circle around the shorter ribbon. **(1 mark)**

2 metres   **100 centimetres**

d. The temperature in Miss Brown's classroom is 18 °C.
In Mr Smith's classroom, the temperature is 20 °C.
Put a circle around the teacher who has the warmer classroom. **(1 mark)**

Miss Brown   **Mr Smith**

Top tip!

Learn these unit facts off by heart:
100 cm = 1 m
1000 g = 1 kg
1000 ml = 1 litre

Total $\frac{}{21}$

1 Reis has 65p. He has a 50p coin and three more coins.

Write the correct amounts on the coins. **(1 mark)**

  ⌀ ⌀ ⌀
....

2 Tara has these coins.

How much has she got altogether? **(1 mark)**

.... p

3 Will needs £1 to pay for a drink. He has four coins.

Write the correct amounts on the coins. **(1 mark)**

⌀ ⌀ ⌀ ⌀
....

4 Dev and Sanjay each have 50p. They have each lost one coin.

Write down the missing coins. **(2 marks)**

Dev   ⌀
....

Sanjay   ⌀
....

5 Add these amounts.

a. 25p + 35p = ... **(1 mark)**

b. 28p + 22p = ... **(1 mark)**

c. 43p + 56p = ... **(1 mark)**

d. 38p + 23p = ... **(1 mark)**

NUMBER AND PLACE VALUE

Counting forwards and back
pages 4–5

1 a. 34, 36, 38 (1 mark)
   b. 33, 36, 39 (1 mark)
   c. 55, 60, 65 (1 mark)
   d. 70, 80, 90 (1 mark)
2 a. 18, 16, 14 (1 mark)
   b. 6, 3, 0 (1 mark)
   c. 15, 10, 5 (1 mark)
   d. 45, 35, 25 (1 mark)
3 a. 37, 43, 47 (1 mark)
   b. 42, 36, 27 (1 mark)
   c. 44, 59, 64 (1 mark)
   d. 62, 42, 2 (1 mark)
4 −5 (1 mark: Accept subtract five.)
5 5 (1 mark: Accept 41, 51, 61, 71, 81.)

6

			42	44	46			52	

(1 mark)

7

	57			66	69	72			

(1 mark)

8

	45	46	47	48	49	
54						
						70

(1 mark)

9 61, 66, 71 (1 mark)

Place value
pages 6–7

1 a. 6 (tens) (1 mark)
   b. 4 (tens) and 8 (units) (1 mark: Both answers needed for 1 mark.)
   c. 3 (units) and 8 (tens) (1 mark: Both answers needed for 1 mark.)
   d. 75 (1 mark)
   e. 94 (1 mark)
2 41 (1 mark)
3 73 (1 mark)
4 62 (1 mark)
5 a. 47 (1 mark)
   b. 94 (1 mark)
   c. 68 (1 mark)

6

	7	
16	17	18
	27	

(4 marks: Award 1 mark for each correct number.)

7 a. 44 (1 mark)
   b. 37 (1 mark)
   c. 9 (1 mark)
8 a. 78 (1 mark)
   b. 38 (1 mark)
9 48 (1 mark)

Numbers and number lines
pages 8–9

1 **a.** 65   (1 mark)
    b. 15   (1 mark)
    c. 60   (1 mark)

2 46   (1 mark)

3 56   (1 mark)

4 **a.** 30   (1 mark)
    b. 50   (1 mark)
    c. 60   (1 mark)
    d. 25   (1 mark)

5 **a.**   **(1 mark: Accept arrow drawn to within 2 mm of line.)**

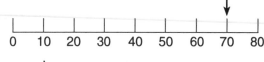

b.   **(1 mark: Accept arrow drawn to within 2 mm of line.)**

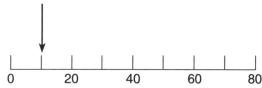

c.   **(1 mark: Accept arrow drawn to within 2 mm of line.)**

d.   **(1 mark: Accept arrow drawn to within 2 mm of line.)**

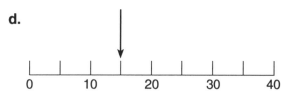

More about numbers
pages 10–11

1 **a.** 35   (1 mark)
    b. 76   (1 mark)
    c. 50   (1 mark)
    d. 83   (1 mark)

2 **a.** twenty   (1 mark)
    b. sixteen   (1 mark)
    c. ninety-two   (1 mark)
    d. sixty-four   (1 mark)
                        (For 2a.–d., accept unambiguous spellings.)

3 19, 43, 56, 70   (1 mark)

4 75, 68, 61, 17   (1 mark)

5 **a.** >   (1 mark)
    b. >   (1 mark)
    c. <   (1 mark)
    d. =   (1 mark)
    e. <   (1 mark)
    f. >   (1 mark)

6 80 ✓ (1 mark)

7 9 ✓ (1 mark)

8 **Accept the following answers:**

57–63 inclusive (1 mark)

65–83 inclusive (1 mark)

85–89 inclusive (1 mark)

9 **Accept the following answers:**

54–70 inclusive (1 mark)

Two numbers between 39–52 inclusive, in descending order (2 marks)

CALCULATIONS

Adding and subtracting

pages 12–13

1 **a.** 21 (1 mark)

b. 61 (1 mark)

c. 44 (1 mark)

d. 82 (1 mark)

2 **a.** 17 (1 mark)

b. 37 (1 mark)

c. 49 (1 mark)

d. 66 (1 mark)

3 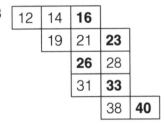 (3 marks: Award 2 marks for 3 or 4 correct numbers. Award 1 mark for 1 or 2 correct numbers.)

4 **a.** 13 (1 mark)

b. 20 (1 mark)

c. 16 (1 mark)

d. 21 (1 mark)

5 **a.** 6 (1 mark)

b. 5 (1 mark)

c. 4 (1 mark)

d. 24 (1 mark)

e. 5 (1 mark)

f. 5 (1 mark)

6 **a.** ✓ (1 mark)

b. ✗ (1 mark)

c. ✓ (1 mark)

d. ✗ (1 mark)

e. ✓ (1 mark)

7 Any pair of numbers that total 19; for example, 19 + 0, 18 + 1, 17 + 2, etc.

(1 mark: Accept correct numbers in either order.)

8 Any three numbers that total 12; for example, 12 + 0 + 0, 11 + 1 + 0, 10 + 2 + 0, etc.

(1 mark: Accept correct numbers in either order.)

Multiplying and dividing

pages 14–15

1 **a.** 12 (1 mark)

b. 20 (1 mark)

c. 70 (1 mark)
d. 16 (1 mark)
e. 15 (1 mark)
f. 45 (1 mark)

2 a. 2 (1 mark)
b. 10 (1 mark)
c. 5 (1 mark)
d. 6 (1 mark)
e. 9 (1 mark)
f. 4 (1 mark)

3 $10 \times 2 = 20$ (1 mark: Accept $2 \times 10 = 20$.)

4 a. 7 (1 mark)
b. 6 (1 mark)
c. 7 (1 mark)
d. 5 (1 mark)
e. 4 (1 mark)
f. 8 (1 mark)

5 a. $6 \times 5 = 30$ (1 mark: Accept $5 \times 6 = 30$.)
b. $2 \times 10 = 20$ (1 mark: Accept $10 \times 2 = 20$.)
c. $8 \times 2 = 16$ (1 mark: Accept $2 \times 8 = 16$.)
d. $2 \times 5 = 10$ (1 mark: Accept $5 \times 2 = 10$.)

6

×	2	5	10
3	6	**15**	**30**
7	**14**	35	**70**
8	**16**	**40**	80

(3 marks: Award 2 marks for four or five correct answers. Award 1 mark for two or three correct answers.)

7

×	2	5	10
4	8	20	**40**
5	**10**	25	50
9	18	**45**	90

(3 marks: Award 2 marks for four or five correct answers. Award 1 mark for two or three correct answers.)

8 a. $8 \times 2 = 16$ (1 mark)
b. $2 \times 8 = 16$ (1 mark)
(Accept answers to a. and b. in either order.)
c. $16 \div 2 = 8$ (1 mark)
d. $16 \div 8 = 2$ (1 mark)
(Accept answers to c. and d. in either order.)

9 a. ✓ (1 mark)
b. ✗ (1 mark)

Problem solving
pages 16–17

1 a. 89 (1 mark)
b. 32 (1 mark)
c. 38 (1 mark)
d. 70 (1 mark)
e. 126 (1 mark)
f. 27 (1 mark)
g. 33 (1 mark)
h. 112 (1 mark)

2 a. 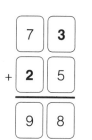 **(1 mark)**

b.

8	4
− 2	8
5	6

(1 mark)

3 45, 51, 13 and 27 should be circled.

(2 marks: Award only 1 mark for two or three correct answers.)

4 a. 13 **(2 marks: Award only 1 mark for the correct method (40 − 27) but the wrong answer.)**

b. 43 **(2 marks: Award only 1 mark for the correct method (28 + 15) but the wrong answer.)**

c. 30 **(2 marks: Award only 1 mark for the correct method (5 × 6) but the wrong answer.)**

d. 4 **(2 marks: Award only 1 mark for the correct method (20 ÷ 5) but the wrong answer.)**

e. 7 **(2 marks: Award only 1 mark for the correct method (35 ÷ 5) but the wrong answer.)**

FRACTIONS

What are fractions?
pages 18–19

1 Fraction   Fraction name   **(2 marks: Award only 1 mark for two or three lines drawn correctly.)**

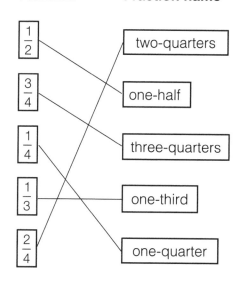

2 a. $\frac{1}{4}$ **(1 mark)**

b. $\frac{1}{2}$ **(1 mark)**

c. $\frac{1}{3}$ **(1 mark)**

d. $\frac{1}{2}$ or $\frac{2}{4}$ **(1 mark)**

3 a. $\frac{1}{2}$ (or equivalent) **(1 mark)**

b. $\frac{1}{2}$ (or equivalent) **(1 mark)**

4 Boxes 1 and 3 ☑   **(2 marks)**
5 Boxes 1 and 4 ☑   **(2 marks)**
6 Box 3 ☑   **(1 mark)**
7 Row 3 ☑   **(1 mark)**

Finding fractions
pages 20–21

1 **a.** Two sections should be shaded.   **(1 mark)**
   b. Two sections should be shaded.   **(1 mark)**
   c. Four sections should be shaded.   **(1 mark)**
   d. Two sections should be shaded.   **(1 mark)**
2 **a.** One section should be shaded.   **(1 mark)**
   b. Six sections should be shaded.   **(1 mark)**
   c. One section should be shaded.   **(1 mark)**
   d. Two sections should be shaded.   **(1 mark)**
3 Three counters should be shaded.   **(1 mark)**
4 Six counters should be shaded.   **(1 mark)**
5 **a.** 6 cm   **(1 mark)**
   b. £2   **(1 mark)**
   c. 5 kg   **(1 mark)**
   d. 15 kg   **(1 mark)**
   e. 5 m   **(1 mark)**
   f. £9   **(1 mark)**
6 **a.** £8   **(2 marks: Award only 1 mark for the correct method (16 ÷ 2) but the wrong answer.)**

   b. 2 kg   **(2 marks: Award only 1 mark for the correct method (8 ÷ 4) but the wrong answer.)**

   c. 10 pages   **(2 marks: Award only 1 mark for the correct method (30 ÷ 3) but the wrong answer.)**

   d. 9 toy bears   **(2 marks: Award only 1 mark for the correct method (12 ÷ 4 and 3 × 3) but the wrong answer.)**

7 **a.** $\frac{1}{4}$ (or equivalent)   **(1 mark)**

   b. $\frac{3}{4}$ (or equivalent)   **(1 mark)**

MEASUREMENT

Units and scales
pages 22–23

1   **(2 marks: Award only 1 mark for two or three lines drawn correctly.)**

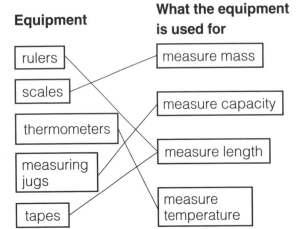

Equipment	What the equipment is used for
rulers	measure mass
scales	measure capacity
thermometers	measure length
measuring jugs	measure temperature
tapes	

2 kilogram ☑   **(1 mark)**
3 litre ☑   **(1 mark)**
4 centimetre ☑   **(1 mark)**
5 8 cm   **(1 mark)**
6 17 °C   **(1 mark)**
7 3 litres   **(1 mark)**
8 700 g   **(1 mark)**
9 300 ml   **(1 mark)**

Comparing measures
pages 24–25

1 rabbit, dog, horse, elephant   **(1 mark)**
2 bath, bucket, bottle, cup   **(1 mark)**
3 bike, car, minibus, lorry   **(1 mark)**
4 150 grams ☑   **(1 mark)**
5 300 millilitres ☑   **(1 mark)**
6 **a.** <   **(1 mark)**
   b. <   **(1 mark)**
   c. >   **(1 mark)**
   d. =   **(1 mark)**
   e. >   **(1 mark)**
   f. <   **(1 mark)**
7 **a.** =   **(1 mark)**
   b. >   **(1 mark)**
   c. >   **(1 mark)**
   d. =   **(1 mark)**
   e. >   **(1 mark)**
   f. =   **(1 mark)**
8 **a.** Lola   **(1 mark)**
   b. 2 litres   **(1 mark)**
   c. 100 centimetres   **(1 mark)**
   d. Mr Smith   **(1 mark)**

Money
pages 26–27

1 5p, 5p, 5p   **(1 mark)**
2 59p   **(1 mark)**
3 50p, 20p, 20p, 10p   **(1 mark: Accept correct coins in any order.)**
4 Dev: 5p   **(1 mark)**
   Sanjay: 20p   **(1 mark)**
5 **a.** 60p   **(1 mark)**
   b. 50p   **(1 mark)**
   c. 99p   **(1 mark)**
   d. 61p   **(1 mark)**
6 **a.** 55p   **(1 mark)**
   b. 25p   **(1 mark)**
   c. 5p   **(1 mark)**
   d. 40p   **(1 mark)**
7 **a.** 70p   **(2 marks: Award only 1 mark for the correct method (£5 – £4.30) but the wrong answer.)**
   b. £1.30   **(2 marks: Award only 1 mark for the correct method (55p + 75p) but the wrong answer. Do not accept £1.3.)**

c. £8.00 or £8

d. 5 coins

(2 marks: Award only 1 mark for the correct method (£10 − £2) but the wrong answer.)
(1 mark: Accept 20p, 20p, 5p, 2p, 2p.)

Time
pages 28–29

1 a. 8 o'clock (1 mark)
  b. half past 4 (1 mark)
  c. quarter to 3 (1 mark)
  d. quarter past 10 (1 mark)
  e. 20 past 7 (1 mark)
  f. 10 to 6 (1 mark)

(For a. to f., accept variations in the presentation of the times, e.g. 8.00, 8:00; $\frac{1}{2}$ past 4, 4.30; twenty past seven, 7.20, etc. Accept unambiguous spellings.)

2 60 minutes (1 mark)

3 The answer should show an understanding that there are 24 hours in a day. It should state:
- there are 24 hours in a day **or**
- there are two lots of 12 hours in a day **or**
- the hour hand goes round twice. (1 mark)

4 a. (1 mark)

b. (1 mark)

c. (1 mark)

d. (1 mark)

e.  (1 mark)

f. (1 mark)

(For a. to f., accept minor inaccuracies in drawing as long as the intention is clear.)

5 30 minutes, 3 hours, 3 days, 3 weeks (1 mark)

6 is shorter than ☑ (1 mark: Accept 'is shorter than' written on answer line.)

GEOMETRY

2D shapes
pages 30–31

1 a. square (1 mark)
  b. triangle (1 mark)
  c. pentagon (1 mark)
  d. rectangle (1 mark)

2 Description   Shape

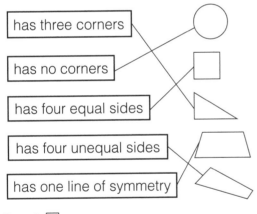

has three corners

has no corners

has four equal sides

has four unequal sides

has one line of symmetry

(2 marks: Award only 1 mark for two or three lines drawn correctly.)

3 Box 3 ☑ (1 mark)

4 a.

(1 mark: Accept minor inaccuracies, but the line must be within 2 mm of vertices. Accept other lines of symmetry, for example:

)

b.

(1 mark: Accept minor inaccuracies, but the line must be within 2 mm of vertices.)

5

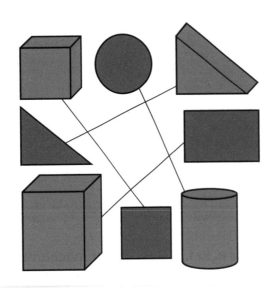

(2 marks: Award only 1 mark for two or three lines drawn correctly. Accept a line drawn from triangular prism to rectangle.)

3D shapes
pages 32–33

1 **a.** cuboid   (1 mark)
   b. cylinder   (1 mark)
   c. pyramid   (1 mark)
   d. cube   (1 mark)
2 cone, cylinder and sphere ☑   (2 marks: Award only 1 mark for two shapes ticked and no incorrect answers.)

3 This shape is a **pyramid**.   (1 mark)
   It has four faces that are **triangles**.   (1 mark)
   It has one face that is a **square**.   (1 mark)
4 **a.** 6, 12, 8   (3 marks: Award 1 mark for each correct number.)
   b. 5, 8, 5   (3 marks: Award 1 mark for each correct number.)
   c. 4, 6, 4   (3 marks: Award 1 mark for each correct number.)
   d. 5, 9, 6   (3 marks: Award 1 mark for each correct number.)
5 **a.** sphere   (1 mark)
   b. cone   (1 mark)
   c. cylinder   (1 mark)
   d. cuboid   (1 mark)

Position and direction
pages 34–35

1 **a.** 40   (1 mark)
   b. 30   (1 mark)
   c. 30   (1 mark)
   d. 50   (1 mark)
2 **a.** Box 1 ☑   (1 mark)
   b. Box 3 ☑   (1 mark)
   c. Box 3 ☑   (1 mark)
3 Box 3 ☑   (1 mark)
4 Box 2 ☑   (1 mark)
5 Box 1 ☑   (1 mark)
6   (2 marks: Award 2 marks for shape correctly drawn. Award 1 mark for shape with the correct dimensions drawn anywhere on the grid.)

STATISTICS

Bar charts
pages 36–37

1 a. 8   (1 mark)
   b. 15   (1 mark)
   c. 13   (1 mark)
   d. 12   (1 mark)
2 a. 25   (1 mark)
   b. Tuesday   (1 mark)
   c. Monday and Wednesday   (1 mark)
   d. 10   (1 mark)
   e. two   (1 mark)

Pictograms
pages 38–39

1 a. 14   (1 mark)
   b. 4   (1 mark)
   c. 2 (Tariq and Mae)   (1 mark)
   d. Sonia and Rory   (2 marks)
2 a. 60 km   (1 mark)
   b. April   (1 mark)
   c. 5 km   (1 mark)
   d. 5 (May, June, July, August, September)   (1 mark)
   e. 70 km   (1 mark)

Tables
pages 40–41

1 a. 7   (1 mark)
   b. Raj   (1 mark)
   c. 3   (1 mark)
   d. 5   (1 mark)
   e. Tom   (1 mark)
2 a. £3.50   (1 mark: Accept 350p; do not accept £3.5)
   b. £1.80   (1 mark: Accept 180p; do not accept £1.8)
3 a. two   (1 mark)
   b. Tuesday and Friday   (1 mark)
   c. Topic   (1 mark)
   d. PE   (1 mark)
4 a. £15 or £15.00   (1 mark: Do not accept £15.0)
   b. £14 or £14.00   (1 mark: Do not accept £14.0)

Tally charts
pages 42–43

1 a. 8   (1 mark)
   b. 12   (1 mark)
   c. 20   (1 mark)
   d. 26   (1 mark)
2 a. 卌 IIII   (1 mark)
   b. 卌 卌 卌 I   (1 mark)
   c. IIII   (1 mark)
   d. 卌 卌 卌 卌 III   (1 mark)

3

Pet	Tally	Total
dog	JHI III	8
cat	JHI JHI I	11
rabbit	IIII	4
hamster	JHI	5

(4 marks: Award 1 mark for each correct total.)

4 **a.** and **b.**

Weather	Tally	Total
☀	JHI III	8
☁	JHI IIII	9
🌧	IIII	4

(6 marks: Award 1 mark for each correct tally and 1 mark for each correct total.)

c. cloudy

(1 mark)

Mixed Practice Questions
pages 44–49

1 53   **(1 mark)**

2 92   **(1 mark)**

3 43   **(1 mark)**

4 50   **(1 mark)**

5 20   **(1 mark)**

6   **(1 mark)**

	44		54	59			74

7 74 ✓   **(1 mark)**

8 $\frac{1}{4}$ (or equivalent)   **(1 mark)**

9 10 red apples   **(2 marks: 15 ÷ 3 = 5; then 15 − 5 = 10 apples; award only 1 mark for the correct method but the wrong answer.)**

10 =   **(1 mark)**

11 £2   **(2 marks: £3.50 + £4.50 = £8.00; £10.00 − £8.00 = £2.00; award only 1 mark for the correct method but the wrong answer.)**

12 a. cylinder   **(1 mark)**

    b. circle   **(1 mark)**

13 triangle   **(1 mark)**

14 quarter past 8 or 8.15   **(1 mark)**

15 a. £30   **(1 mark)**

    b. £15   **(2 marks)**

16 7   **(1 mark)**

17 a. 71   **(1 mark)**

    b. 45   **(1 mark)**

18 37   **(2 marks: award only 1 mark for the correct method (60 − 23) but the wrong answer.)**

19 20p ✓ 10p ✓ 5p ✓ 2p ✓ 2p ✓

20 9 edges   **(1 mark)**

    6 vertices   **(1 mark)**

    5 faces   **(1 mark)**

6 Subtract these amounts.

a. 60p – 5p = ... **(1 mark)**   **c.** 31p – 26p = ... **(1 mark)**

b. 48p – 23p = ... **(1 mark)**   **d.** 85p – 45p = ... **(1 mark)**

7 Solve these problems. Show your working.

a. George has a £5 note. He spends £4.30.
How much money does he have left?
Give your answer in pence.   **(2 marks)**

.... p

b. Una buys a drink for 55p and a pen for 75p.
How much does she spend altogether?
Give your answer in pounds.   **(2 marks)**

£ ....

c. Sarah has a £10 note. She spends £2 on a loaf of bread.
How much change does she get?
Give your answer in pounds.   **(2 marks)**

£ ....

d. What is the fewest number of coins needed to make 49p?   **(1 mark)**

....

Top tip! When you are doing money sums, only use coins that are used in real life. For example, 8p cannot be made from a 5p coin and a 3p coin because there is no such thing as a 3p coin.

Total $\frac{}{20}$

1 What time is shown on these clocks?

a.

.... **(1 mark)**

d.

.... **(1 mark)**

b.

.... **(1 mark)**

e.

.... **(1 mark)**

c.

.... **(1 mark)**

f.

.... **(1 mark)**

2 It takes James exactly 1 hour to get home.
How many minutes does it take him to get home?   **(1 mark)**

.... minutes

3 Sara says that there are 12 hours on a clock face, so there must be 12 hours in a day.

Why is Sara wrong?   **(1 mark)**

....

....

....

4 Draw hands on each clock face to show the time. **(6 marks)**

a. 6 o'clock

d. 10 past 11

b. quarter to 5

e. quarter past 12

c. half past 9

f. 20 to 1

5 Write these times in order, shortest first: **(1 mark)**

3 hours   **3 days**   **30 minutes**   **3 weeks**

....

6 Complete this statement by ticking (✓) the missing words. **(1 mark)**

45 minutes .... 1 hour.

is longer than   is the same as   is shorter than

Explain that a clock face has two scales: one for hours and one for minutes. Practise telling the time regularly.

Parent tip!

Total $\frac{\quad}{16}$

1 Write the name of each shape. **(4 marks)**

a.

....

b.

....

c.

....

d.

....

2 Draw lines from the descriptions to the shapes. **(2 marks)**
The first one has been done for you.

Description   **Shape**

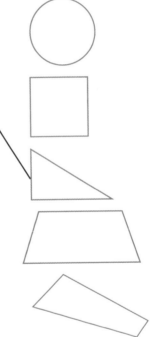

| has three corners |
| has no corners |
| has four equal sides |
| has four unequal sides |
| has one line of symmetry |

3 Tick (✓) the shape where the dotted line is a line of symmetry. **(1 mark)**

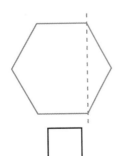 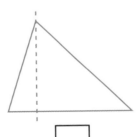  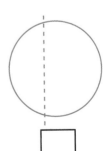

4 Draw a line of symmetry on these shapes. **(2 marks)**

a.

b.

Geometry uses lots of terminology: for example, **equal**, **symmetry**, **triangle**, **3D**. Make sure your child knows what they all mean.

Parent tip!

5 Draw lines from each 3D shape to one of its 2D faces. **(2 marks)**

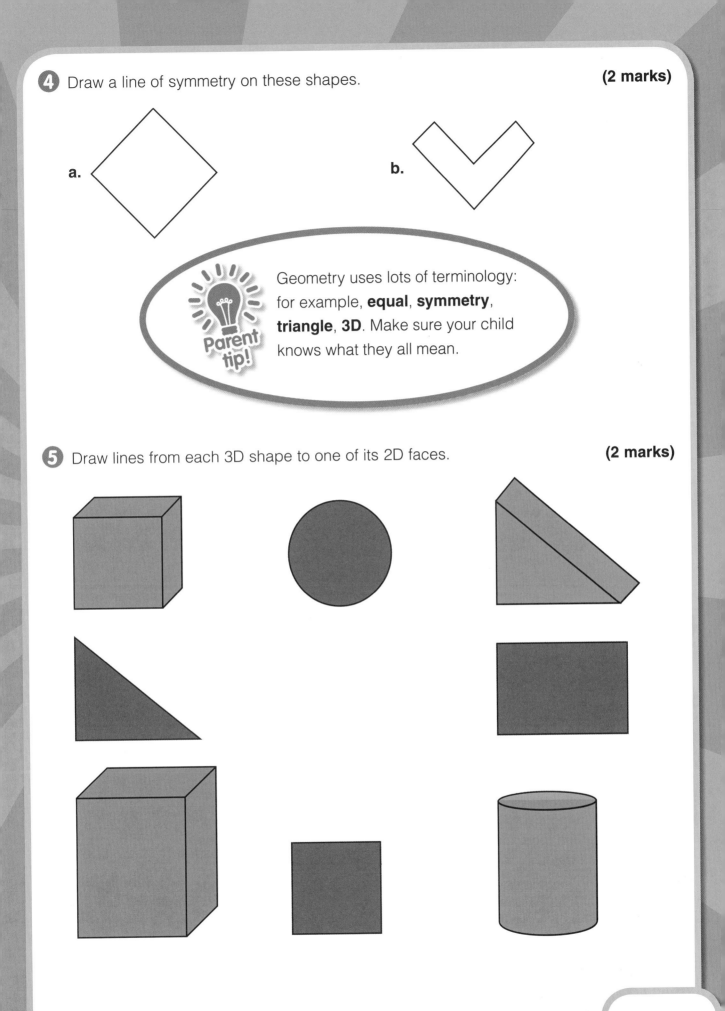

Total — 11

① These drawings show 3D shapes. Write the name of each shape.

(4 marks)

a.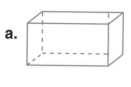

....

b.

....

c.

....

d.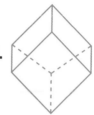

....

② Tick (✓) the names of the 3D shapes that have a curved surface.

(2 marks)

cone ☐   cube ☐   cuboid ☐

cylinder ☐   pyramid ☐   sphere ☐

③ Fill in the missing words.   **(3 marks)**

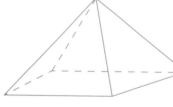

This shape is a ....

It has four faces that are ....

It has one face that is a ....

④ Fill in the missing numbers.

a.

This shape has:

- .... faces   **(1 mark)**
- .... edges   **(1 mark)**
- .... vertices.   **(1 mark)**

b.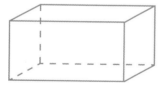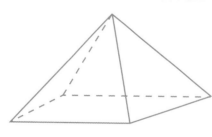

This shape has:

- .... faces   **(1 mark)**
- .... edges   **(1 mark)**
- .... vertices.   **(1 mark)**

c.

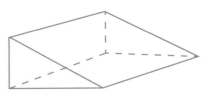

This shape has:

- .... faces **(1 mark)**

- .... edges **(1 mark)**

- .... vertices. **(1 mark)**

d.

This shape has:

- .... faces **(1 mark)**

- .... edges **(1 mark)**

- .... vertices. **(1 mark)**

5 What is the name of the 3D shape that each object is most like? **(4 marks)**

a.  ....

c.  ....

b.  ....

d.  ....

Let your child experiment with 3D shapes. For instance, help them cut up a cereal box to see the 2D shapes it is made from, or count the number of corners and edges it has.

Parent tip!

Total $\frac{}{25}$

1 Write the next number in each sequence. **(4 marks)**

    a. 15   20   25   30   35   ....

    b. 20   22   24   26   28   ....

    c. 80   70   60   50   40   ....

    d. 100   90   80   70   60   ....

2 Tick (✓) the shape that comes next in each sequence. **(3 marks)**

a.

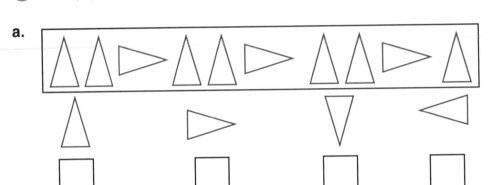

b.

c.

3 This shape is rotated a quarter turn clockwise.

Tick (✓) the new position of the shape. **(1 mark)**

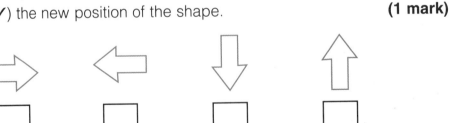

4 Here is a shape.

When it is rotated, its new position is:

Tick (✓) the rotation.   **(1 mark)**

full turn clockwise $\frac{1}{2}$ turn clockwise $\frac{1}{4}$ turn clockwise $\frac{1}{4}$ turn anti-clockwise

☐   ☐   ☐   ☐

5 Here is a shape.

When it is rotated, its new position is:

Tick (✓) the rotation.   **(1 mark)**

90° clockwise   180° clockwise   270° clockwise   360° clockwise

☐   ☐   ☐   ☐

6 Draw the triangle after it has moved 3 squares left and 1 square down.   **(2 marks)**

Top tip!

When you are doing rotations, try using tracing paper to draw over the shape. Then move the shape on the tracing paper, so you can see how it has rotated.

Total $\frac{}{12}$

① This bar chart shows how children travel to school.

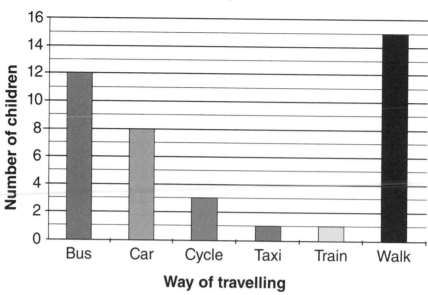

Travelling to school

Number of children

Way of travelling

Answer these questions about the bar chart.

a. How many children travel by car?　　　　　　　　　　　**(1 mark)**

.... children

b. How many children walk to school?　　　　　　　　　　**(1 mark)**

.... children

c. How many children travel by bus and taxi altogether?　　**(1 mark)**

.... children

d. How many more children walk than cycle?　　　　　　　**(1 mark)**

.... children

Top tip!

Always read charts carefully
before you answer questions.
Make sure you know what
the chart is telling you.

2 This bar chart shows the number of children who have breakfast at school.

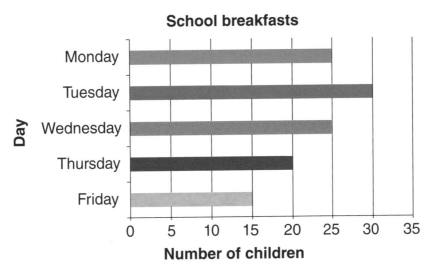

School breakfasts

Answer these questions about the bar chart.

a. How many children had a school breakfast on Monday?　　　　　**(1 mark)**

.... children

b. On what day did 30 children have breakfast at school?　　　　**(1 mark)**

....

c. On which days did the same number of children have breakfast at school?

.... **(1 mark)**

d. How many more children had breakfast on Wednesday than on Friday?　　**(1 mark)**

.... children

e. On how many days were fewer than 25 breakfasts served?　　　　**(1 mark)**

.... days

Total — 9

1 This pictogram shows the number of books that some children read in a month.

 stands for 2 books.

Reading books	
Name	**Number of books read**
Sonia	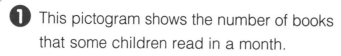
Eliot	
Tariq	
Elijah	
Mae	
Rory	

Answer these questions about the pictogram.

a. How many books did Eliot read? **(1 mark)**

.... books

b. How many more books did Rory read than Tariq? **(1 mark)**

.... books

c. How many children read fewer than 10 books? **(1 mark)**

.... children

d. Name the children who read 12 books. **(2 marks)**

....

Top tip! Always check what each symbol stands for in a pictogram. It might be more than one of something.

2 Ben goes cycling once a month.

This pictogram shows the number of kilometres he cycles.

🚲 stands for 10 kilometres.

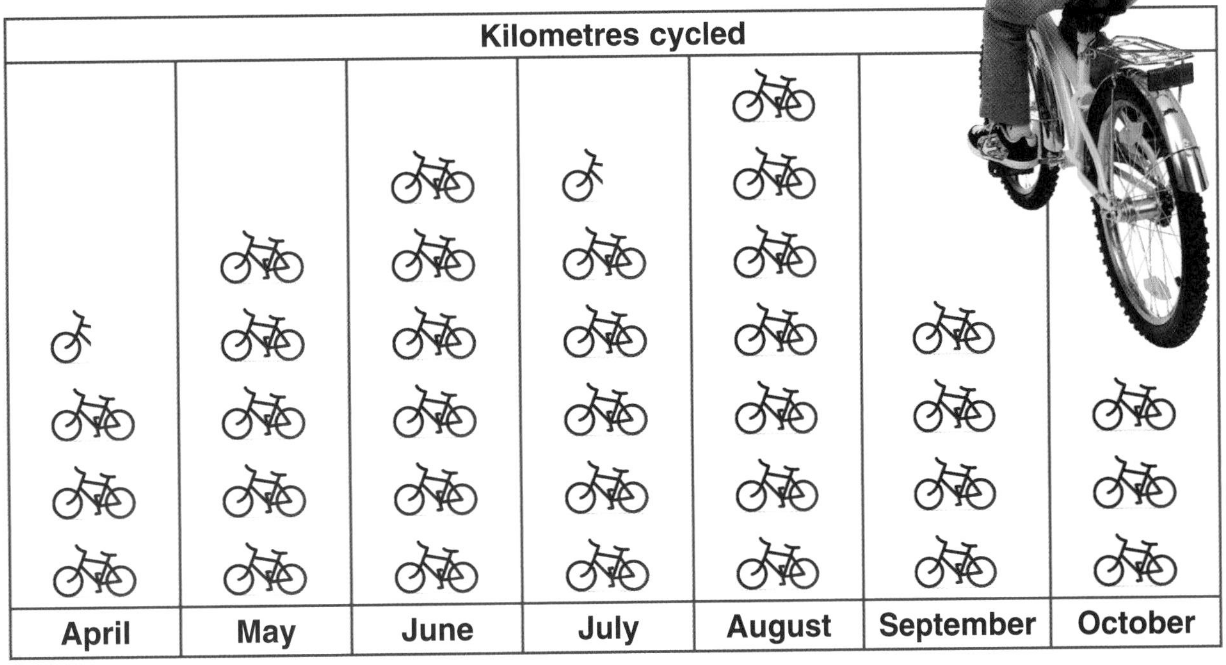

Kilometres cycled

April	May	June	July	August	September	October

Answer these questions about the pictogram.

a. How many kilometres did Ben cycle in June? **(1 mark)**

.... km

b. In which month did Ben cycle 35 kilometres? **(1 mark)**

....

c. How much further did Ben cycle in June than in July? **(1 mark)**

.... km

d. How many months did Ben cycle 40 or more kilometres? **(1 mark)**

.... months

e. What was the longest distance Ben cycled in one month? **(1 mark)**

.... km

Total —— 10

1 This table gives information about some pupils.

Name	Gender	Address	Phone number	Age
Tom	boy	3 Wood Street	532523	7
Raj	boy	15 Ash Terrace	532247	7
Sara	girl	21 Park Grove	543978	7
Natalie	girl	7 Dean Street	554078	7
Ali	boy	11 London Road	521654	6
Tara	girl	5 Wood Street	533867	7

Answer these questions about the table.

a. How old is Sara? **(1 mark)**

.... years old

b. Who lives in Ash Terrace? **(1 mark)**

....

c. How many girls are in the table? **(1 mark)**

.... girls

d. How many children are aged 7? **(1 mark)**

.... children

e. Which boy lives in Wood Street? **(1 mark)**

....

2 This menu gives the prices of some drinks.

Ken's café	
Tea	£1.50
Coffee	£2
Juice	80p
Cola	70p
Lemonade	50p

a. How much does it cost for a cup of coffee and a cup of tea?

.... **(1 mark)**

b. How much does it cost for two lemonades and a juice?

.... **(1 mark)**

3 This is Fay's class timetable for the week.

	Lesson 1	Lesson 2		Lesson 3	Lesson 4
Monday	Maths	English		Topic	Science
Tuesday	English	Maths	**LUNCH**	PE	Art
Wednesday	Maths	English		Topic	Music
Thursday	English	Maths		Science	PE
Friday	Maths	English		Art	Topic

Answer these questions about the timetable.

a. How many PE lessons does Fay have? .... **(1 mark)**

b. On which days does Fay have art? **(1 mark)**

....

c. What lesson does Fay do just before music? **(1 mark)**

....

d. What does Fay do in Lesson 3 on Tuesday? **(1 mark)**

....

4 This poster gives the entry prices for a museum.

Town museum

Adult   £6.00
Child   £3.00

Family   £14.00
(2 adults and 2 children)

a. What is the cost for two adults and a child?

£................ **(1 mark)**

b. Bill is an adult. He pays for his entry with £20. How much change does he get?

£................ **(1 mark)**

Total $\frac{}{13}$

1 Write the numbers these tallies show. **(4 marks)**

a. |||| ||| = ...

b. |||| |||| || = ...

c. |||| |||| |||| |||| = ...

d. |||| |||| |||| |||| |||| | = ...

2 Write a tally for each number. **(4 marks)**

a. 9 = ...

b. 16 = ...

c. 4 = ...

d. 23 = ...

3 Dan asked his friends about their pets.
He made a table of the information.
Complete the total column. **(4 marks)**

Pet	Tally	Total
dog	|||| |||	
cat	|||| ||||		
rabbit	||||	
hamster	||||	

Top tip! Practise counting in fives to count up tallies quickly. Don't forget to add on the extra lines in ones.

4 These are weather signs.

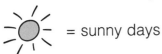 = sunny days

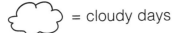 = cloudy days

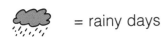 = rainy days

These signs show the weather for the last three weeks:

Complete a tally chart of the weather.

a. Fill in the tallies. **(3 marks)**

b. Fill in the totals. **(3 marks)**

Weather	Tally	Total
☀		
☁		
🌧		

c. Complete this sentence.

There were more .... days than any other. **(1 mark)**

Total $\frac{}{19}$

1

45 + 8 =

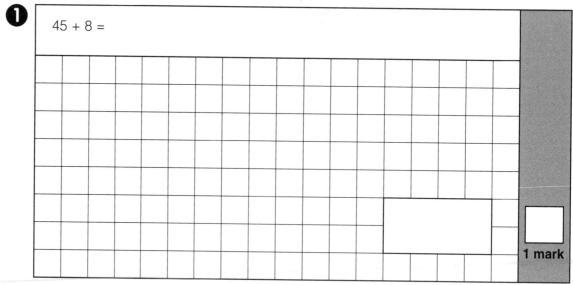

1 mark

2

63 + 29 =

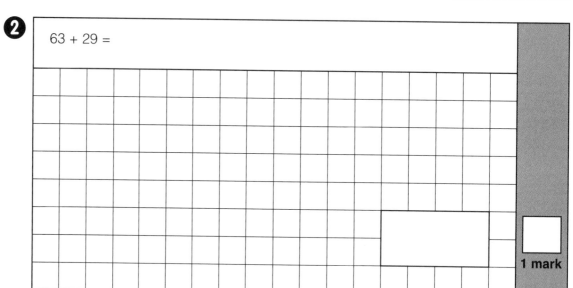

1 mark

3

78 − 35 =

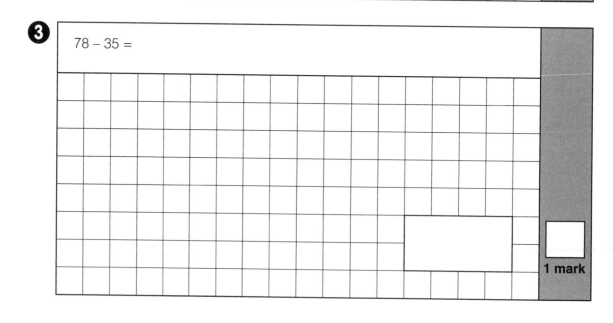

1 mark

4

98 – 48 =

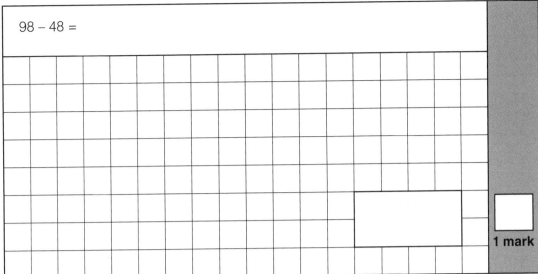

1 mark

 5

6 + 9 + 5 =

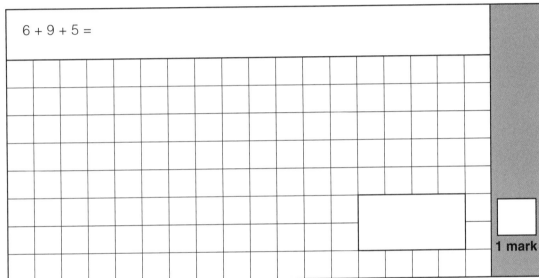

1 mark

6 Here is part of a number line.

	44		54	59			

Write 74 in the correct place. **(1 mark)**

7 Isaac thinks of a number. It has 4 units and 7 tens.

Tick (✓) Isaac's number. **(1 mark)**

 44   47   74   77

8 Here are some stars.

What fraction of the stars is yellow? **(1 mark)**

9 Leo has 15 apples.

$\frac{1}{3}$ of the apples are green. The rest are red.

How many **red** apples does Leo have? **(2 marks)**

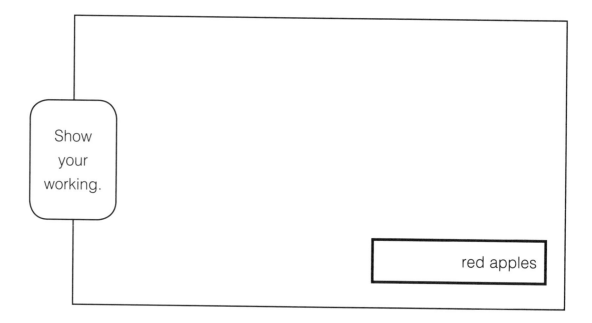

Show your working.

red apples

10 Insert **<**, **>** or **=** to compare these lengths. **(1 mark)**

100 cm ◯ 1m

11 Daisy buys some pens for £4.50 and a book for £3.50
She uses a £10 note.
How much change does she get? **(2 marks)**

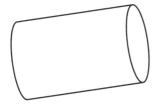
Show your working.

£ _____

12 Look at this shape.

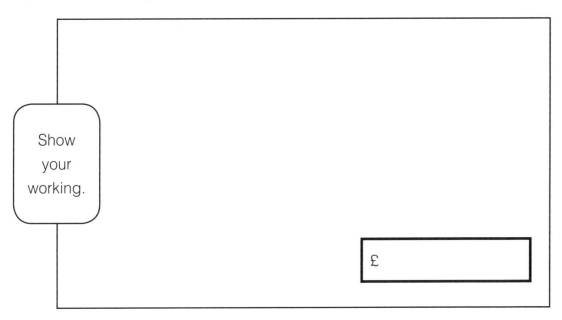

a. What is the name of this shape? **(1 mark)**

b. The shape has two faces that are the same. What shape are they? **(1 mark)**

13 Nadine says that she has a shape that has three corners and three sides.
What is the name of Nadine's shape? **(1 mark)**

14 What time is shown on the clock? **(1 mark)**

15 A school had a fete.

The pictogram shows how much money some stalls raised.

£ stands for £5

Stall	Money raised
cake stall	£ £ £ £ £ £
book stall	£ £ £ £ £
toy stall	£ £ £ £
game stall	£ £ £ £ £ £ £
drinks stall	£ £ £

a. How much money did the cake stall raise? **(1 mark)**

£

b. How much more money did the game stall raise than the toy stall? **(2 marks)**

£

16 What is the missing number? **(1 mark)**

[ ] × 5 = 35

17 a. Write seventy-one as a number. **(1 mark)**

b. Write forty-five as a number. **(1 mark)**

18 Nisha has a box with 60 red and yellow flowers.

23 of the flowers are red. How many of the flowers are yellow? **(2 marks)**

19 Ben has these coins.

(50p) (50p) (20p) (10p) (5p) (5p) (2p) (2p) (1p)

He buys an apple for 39p and gets **no** change.

Tick ✓ the coins he uses. **(1 mark)**

20 Fill in the missing numbers.

This shape has:

edges **(1 mark)**

vertices **(1 mark)**

faces **(1 mark)**

Glossary

2D – Two dimensional; usually that means having a length and a width. 2D shapes are flat and cannot be picked up and handled

3D – Three dimensional; usually that means having a length, width and height. 3D shapes can be picked up and handled

Abacus – A tool for displaying numbers

Adding – Combining two or more numbers to give a total or a sum

Adding words – Words that show you may have to add, such as 'altogether', 'extra', 'more', 'plus', 'sum' and 'total'

Anti-clockwise – To turn in the opposite direction to the hands of a clock

Bar chart – A bar chart shows information as a picture. It uses bars or blocks on a graph

Clockwise – To turn in the same direction as the hands of a clock

Column – The information that runs up and down (vertical) in a table

Counting back – Counting numbers in reverse order, in a group (such as ones, twos, fives), so that the numbers get smaller

Counting forwards – Counting numbers in order, in a group (such as ones, twos, fives), so that the numbers get larger

Denominator – The bottom number in a fraction that shows the number of parts in the whole

Digit – A number from 0–9 that can be used to make other numbers

Dividing – To split or share a number equally

Division words – Words that show you may have to divide, such as 'each', 'every', 'half', 'quarter', 'share', 'split' and 'third'

Edge – A line where two faces meet

Even number – A number that can be divided exactly by 2

Face – A side of a 3D shape

Fraction – A number that shows the part or parts of one whole

Hour – A unit of time; 1 hour is the same as 60 minutes

Metric units – Measures based on groups of ten

Minute – A unit of time; 60 minutes make 1 hour

Multiple – The answer when multiplying whole numbers, e.g. $5 \times 10 = 50$. This makes 50 a multiple of 5 and of 10

Multiplication words – Words that show you may have to multiply, such as 'by', 'double', 'lots', 'times' and 'twice'

Multiplying – Counting or adding in steps of the same number

Number line – A number line shows the position of a number. The numbers can be listed in different ways, such as ones, twos, fives, tens, and so on

Numerator – The top number in a fraction that shows the number of parts you have

Odd number – A number that cannot be divided exactly by 2

Partitioning – Splitting numbers using place value

Pattern – Shapes or objects arranged to follow a rule

Pence / pennies – A unit of money. 100 pennies make £1

Pictogram – A pictogram is a chart. It uses pictures or symbols to stand for a number

Place value – What a digit is worth. This depends on its position in a number

Pounds – A unit of money. 100 pennies make £1.

Rectangle – A four-sided shape that has four right angles

Right angle – A quarter turn. Four quarter turns make a full turn

Rotate – To turn a shape or object around a fixed point

Row – The information that runs across a table (horizontal)

Rule – Tells you how numbers or objects are arranged in a sequence or pattern

Scale – The way the numbers are spread out on the axis of a graph (a line that is used to build a bar chart)

Sequence – Numbers arranged to follow a rule

Subtracting – Taking one number from another to leave a difference

Subtraction words – Words that show you may have to subtract, such as 'difference', 'fewer', 'left', 'less than', 'minus', 'reduce' and 'take away'

Symbols – Symbols are signs that are used instead of words

Tally – Recording what you count by making marks; you count in fives

Vertex – A point or corner where two edges meet

Vertices – More than one vertex